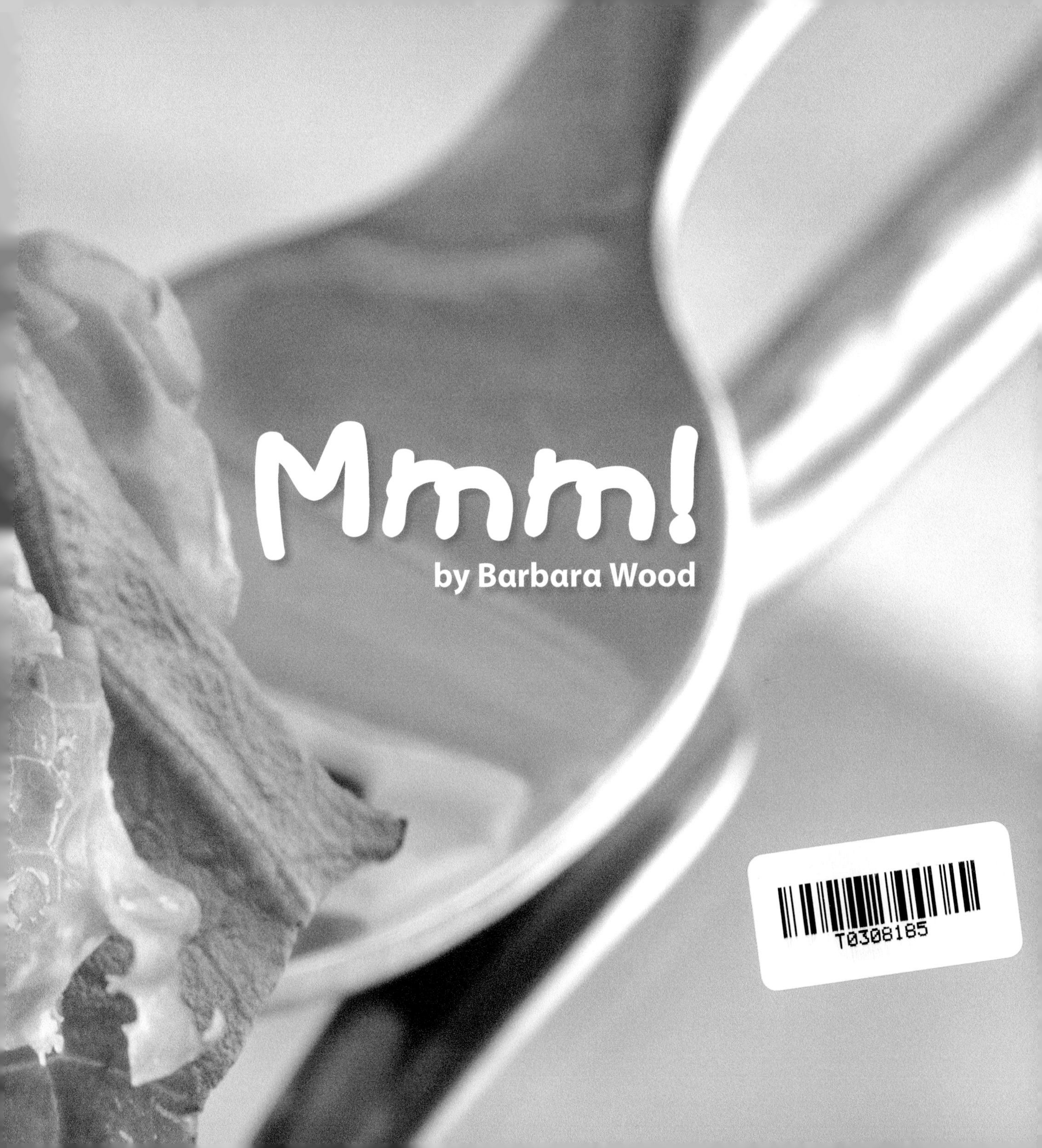

Mmm!

by Barbara Wood

lettuce

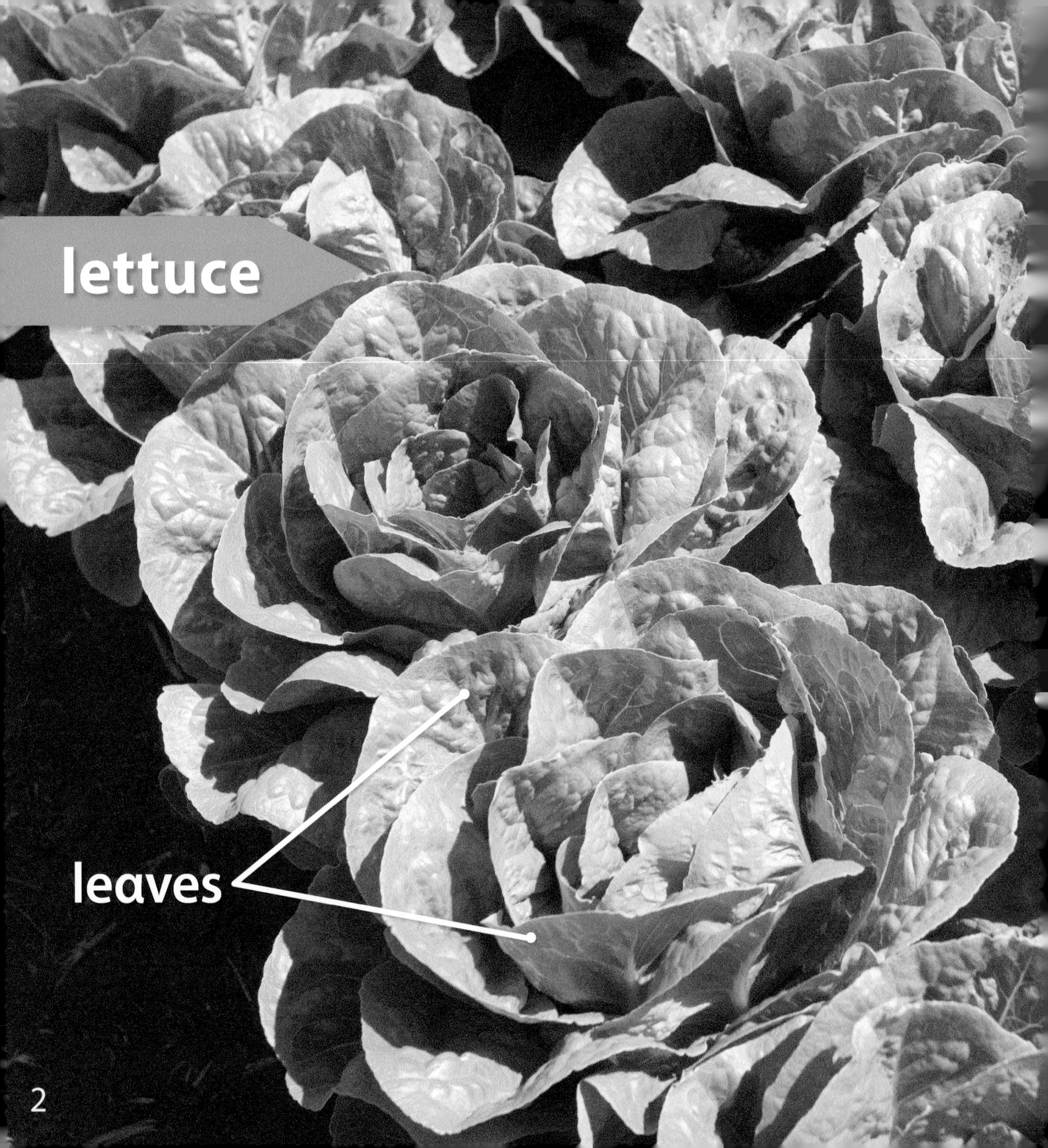

Mmm!

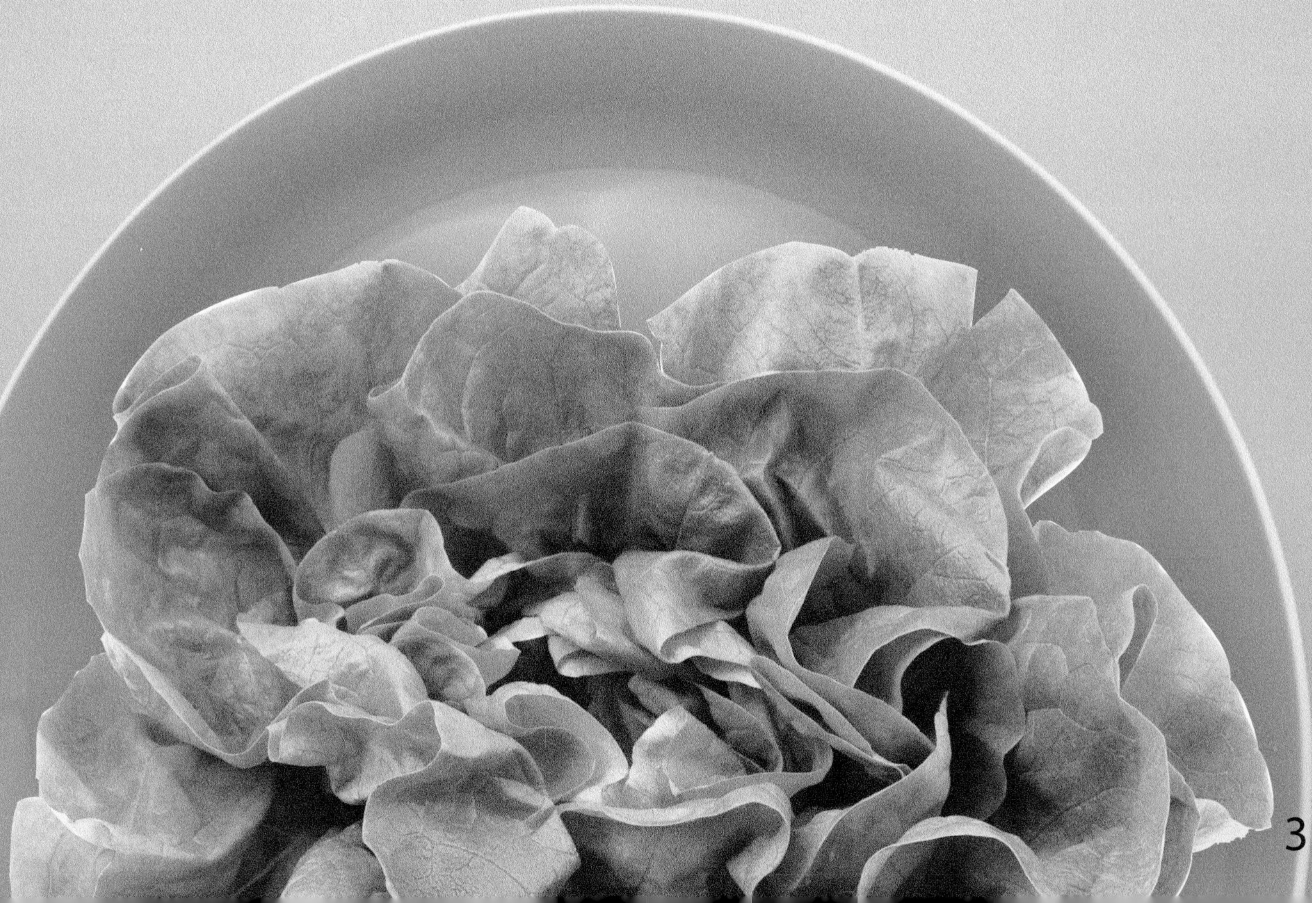

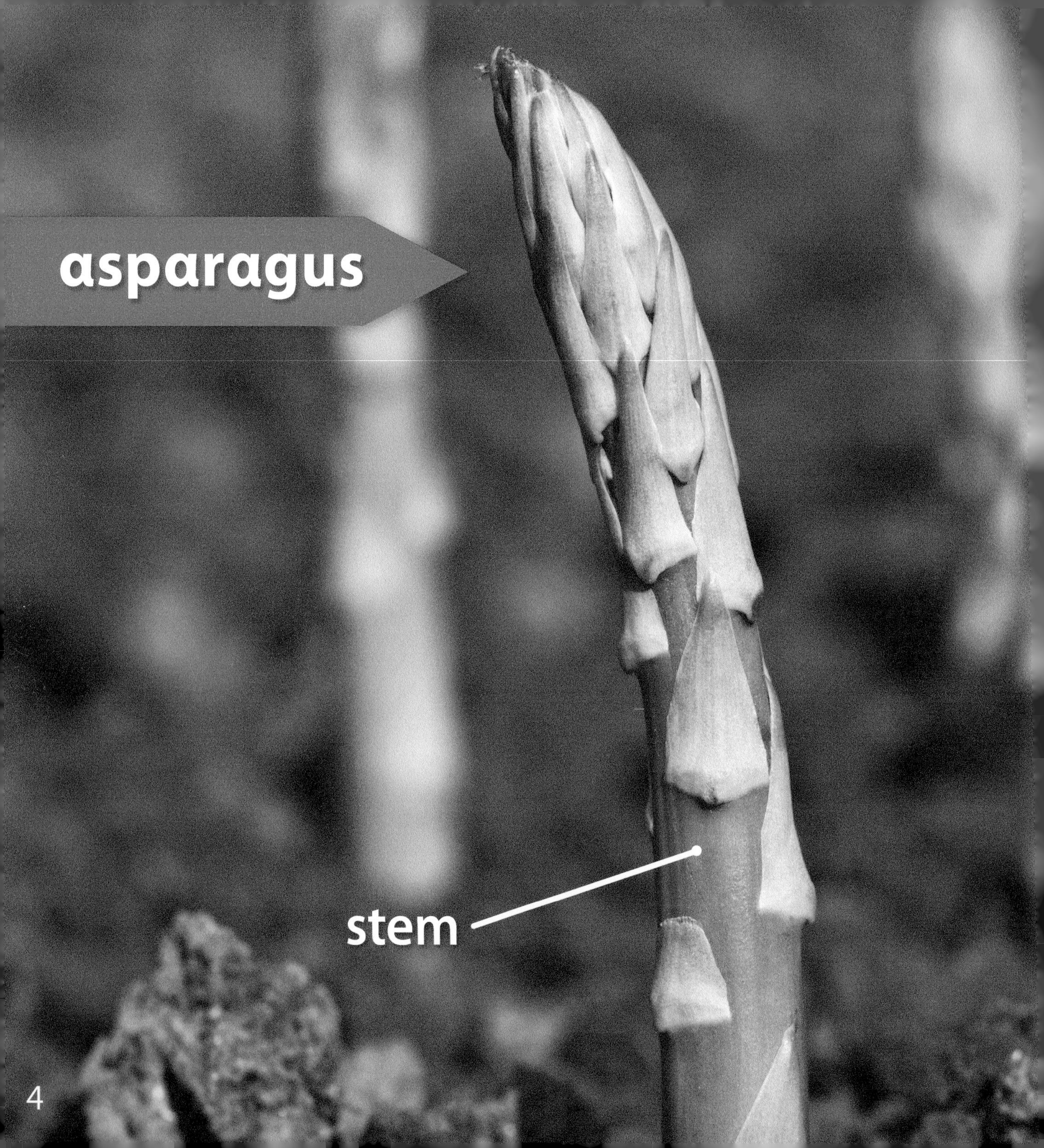
asparagus
stem

Mmm!

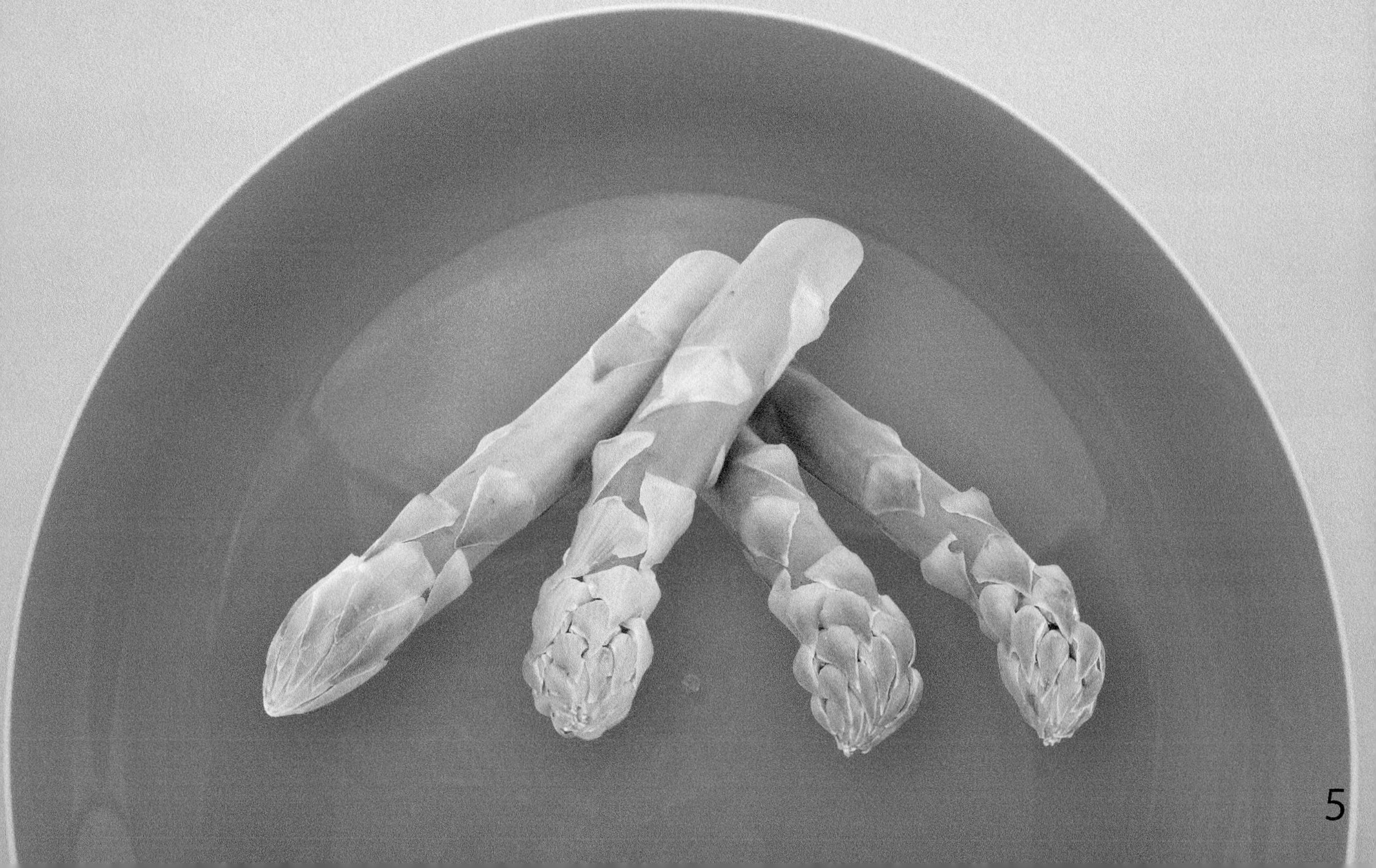

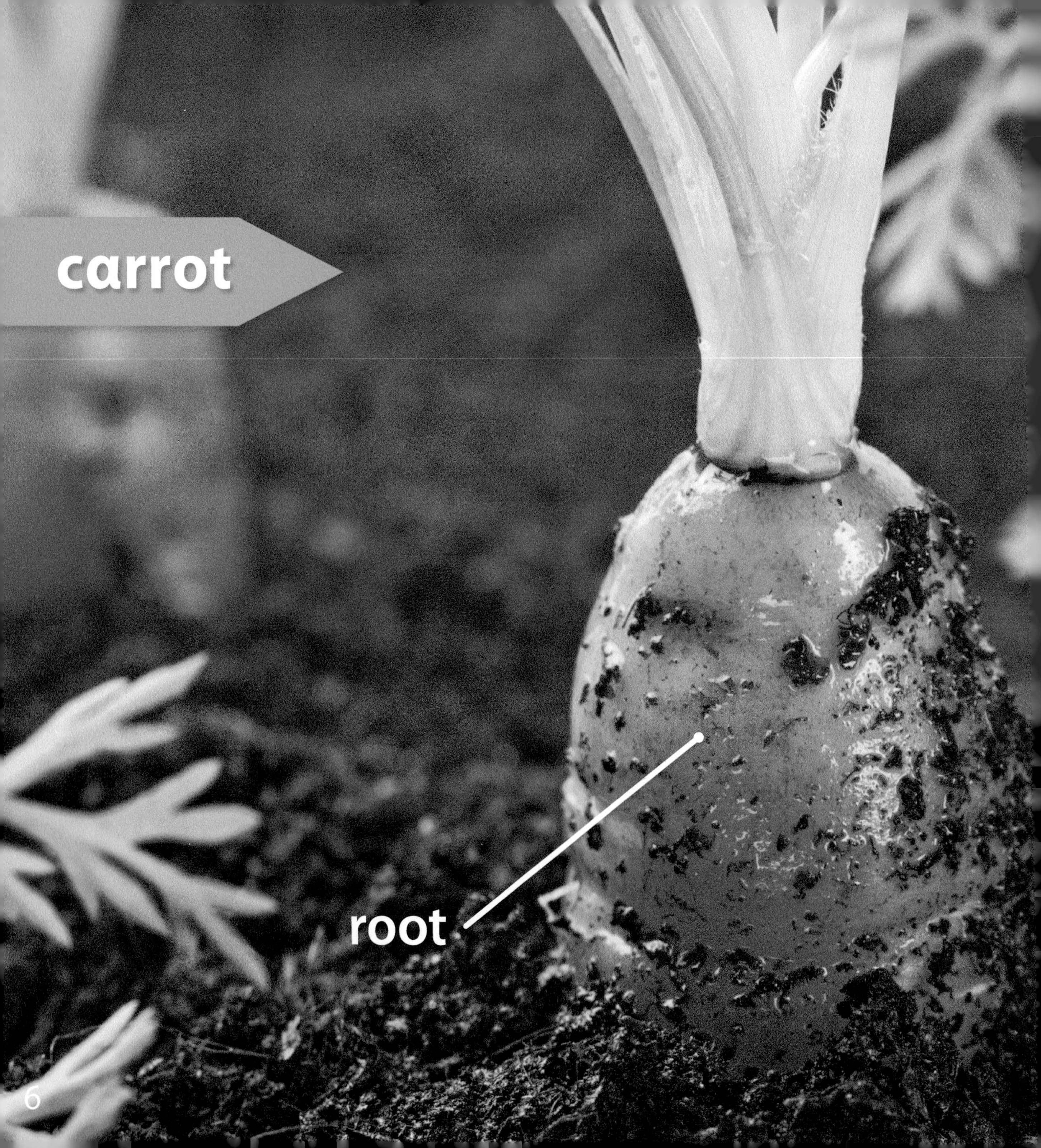
carrot
root

Mmm!

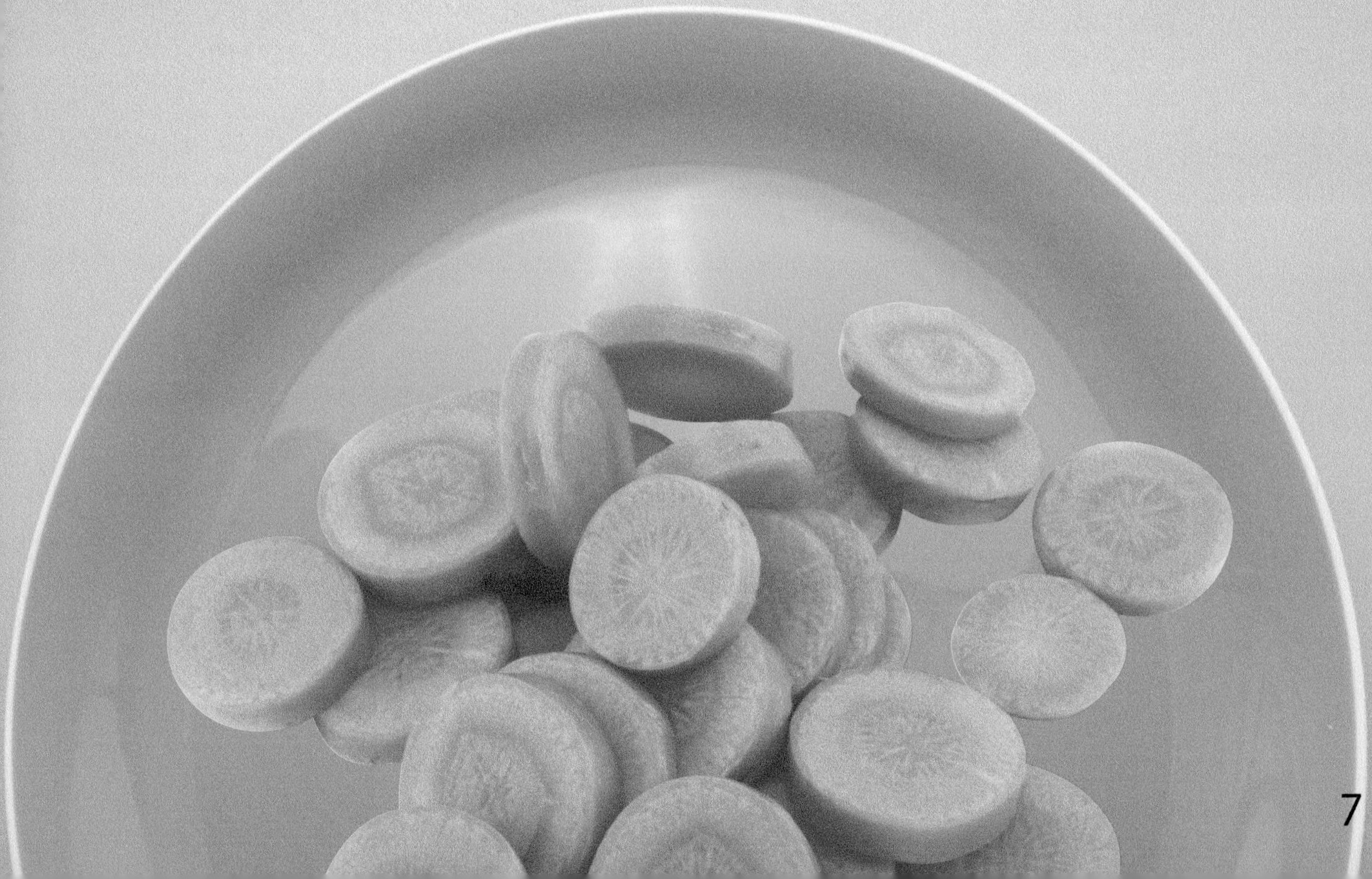

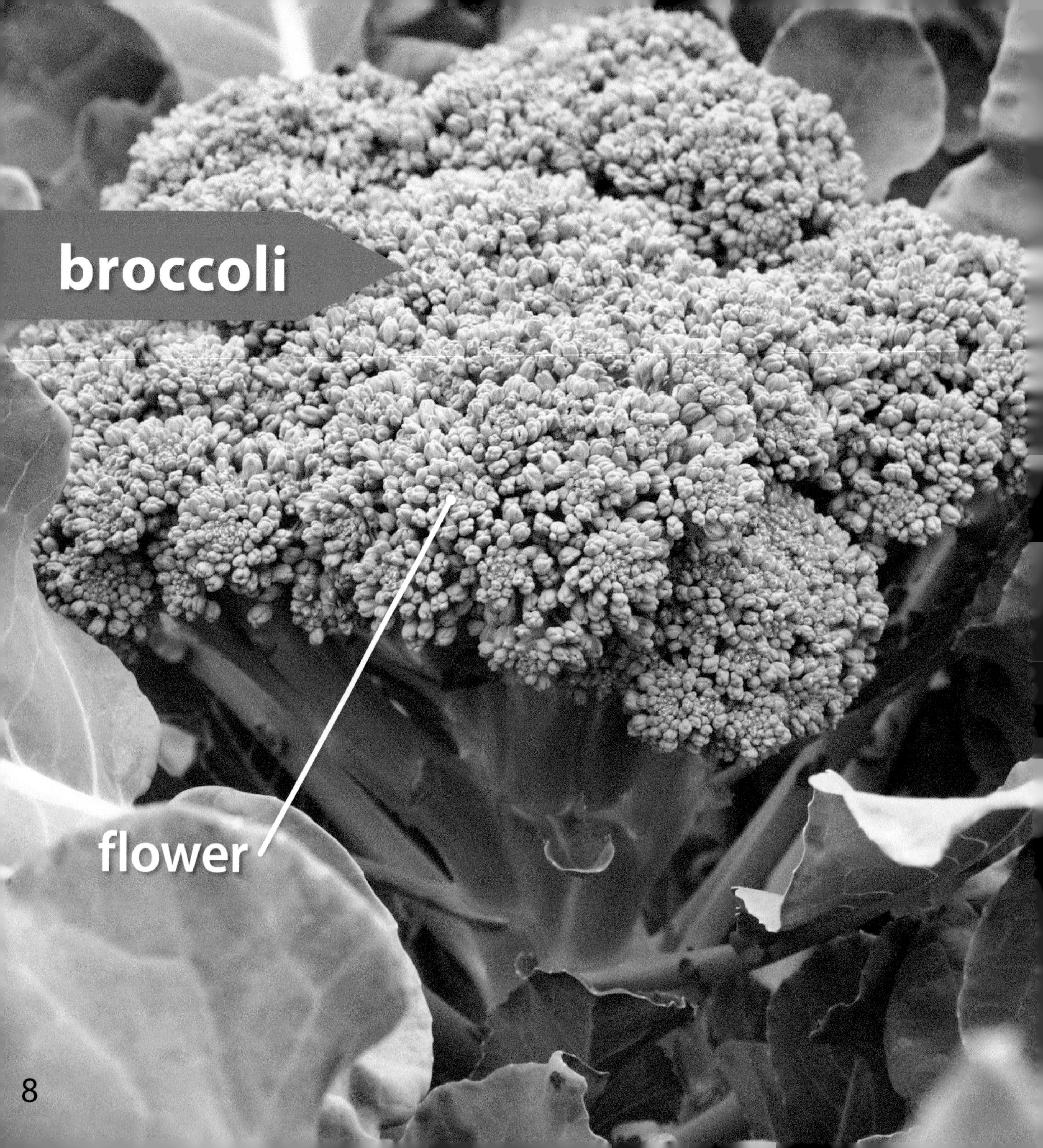
broccoli
flower

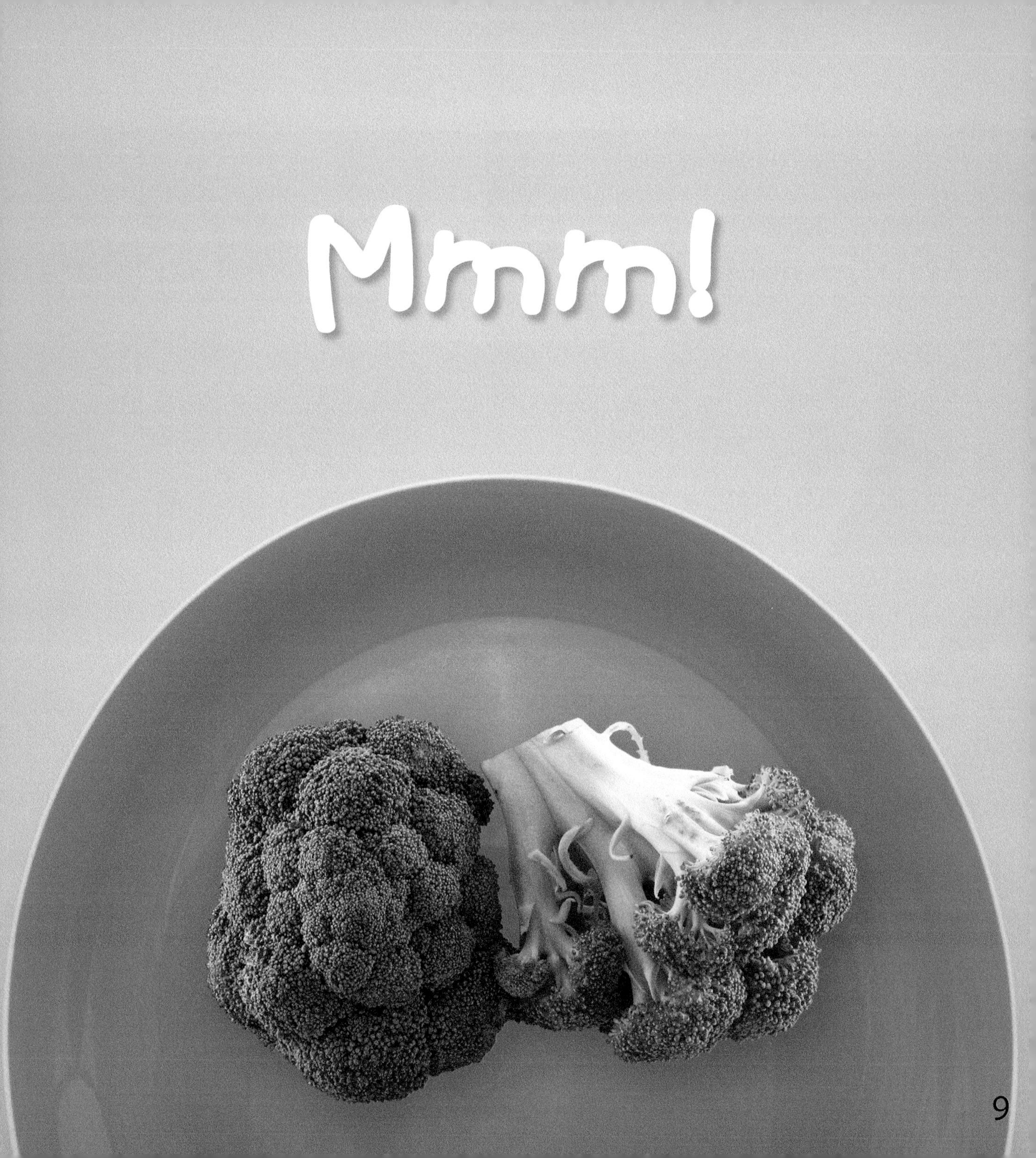
Mmm!

flower

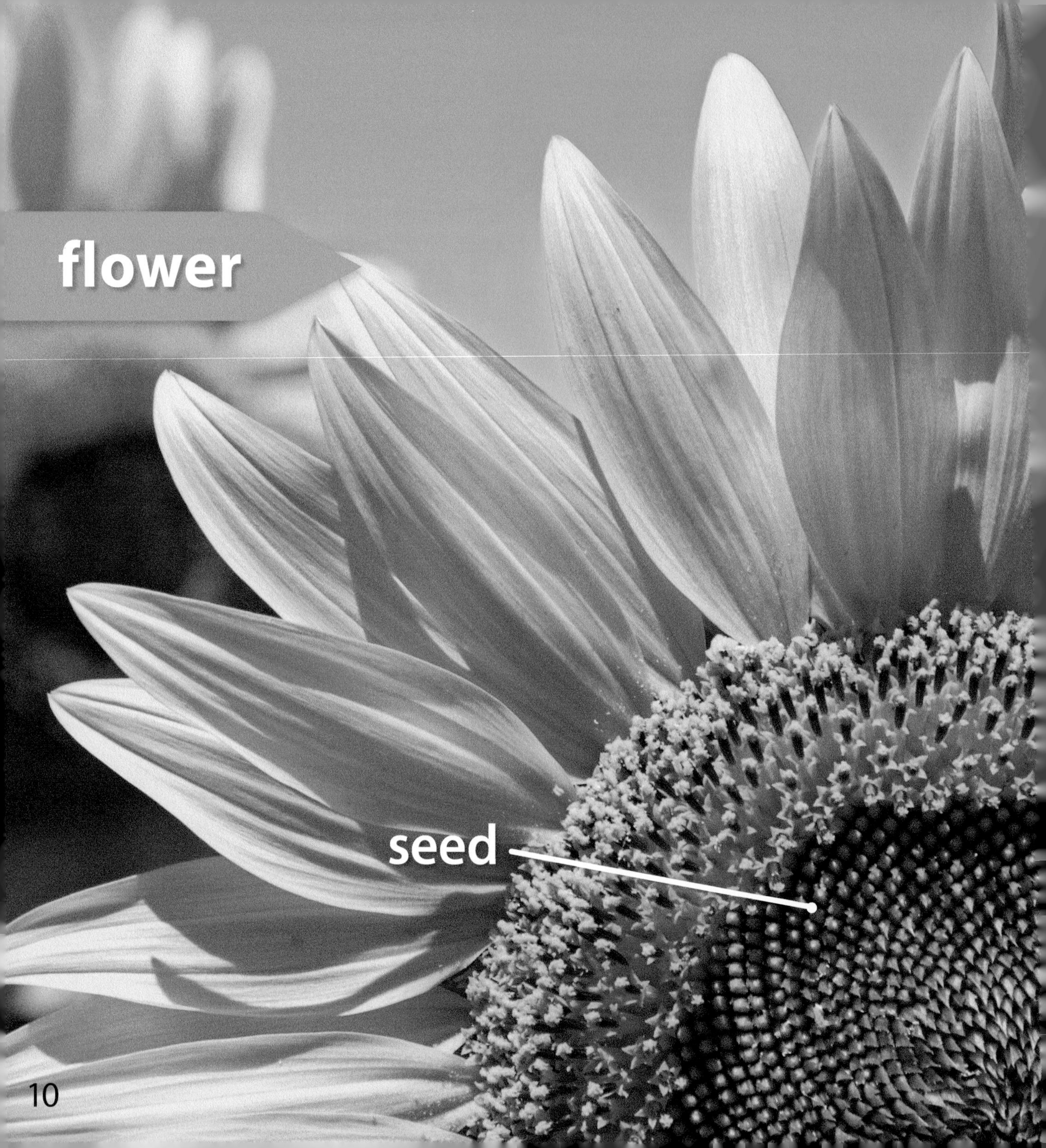

Mmm!

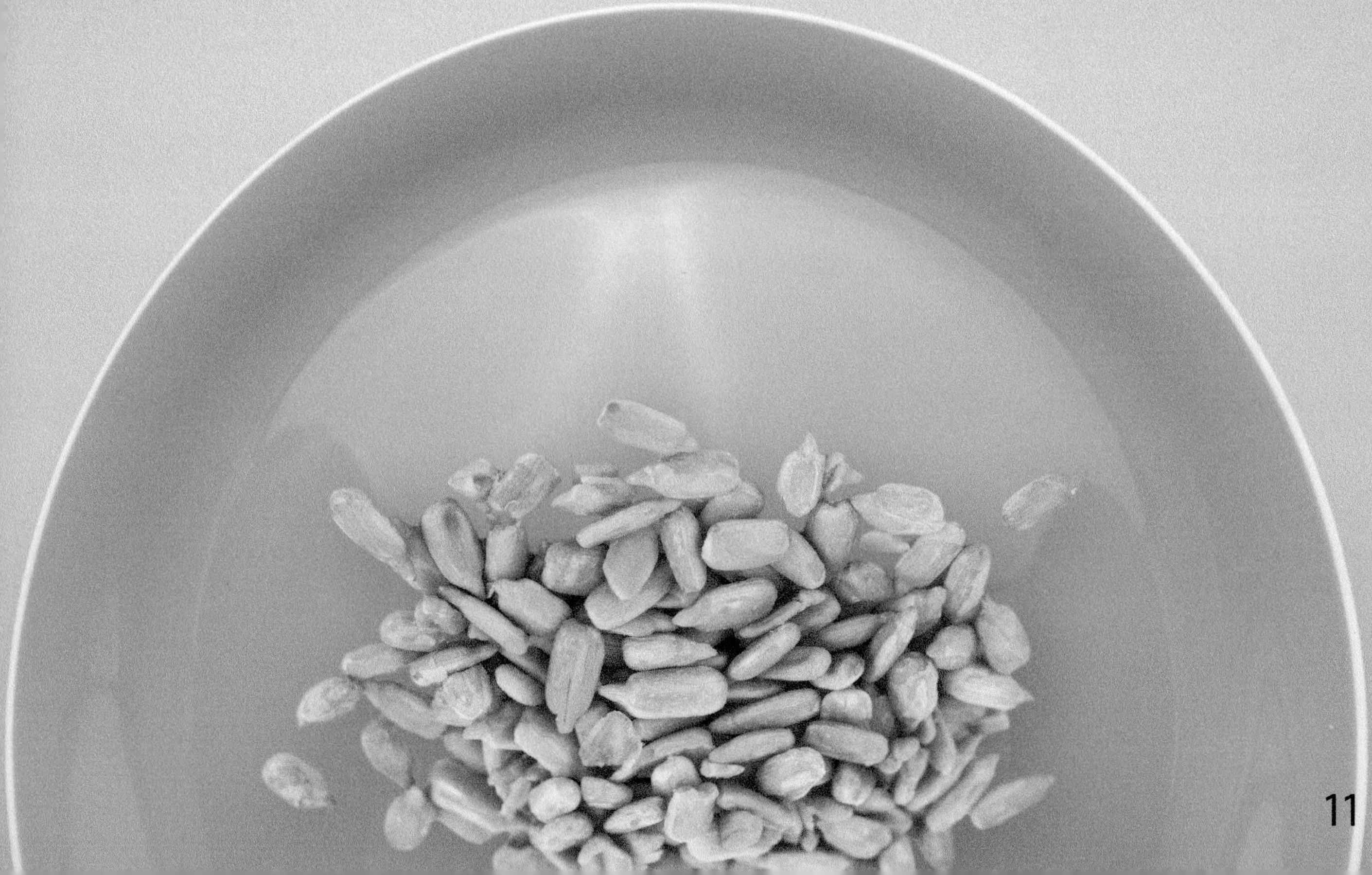

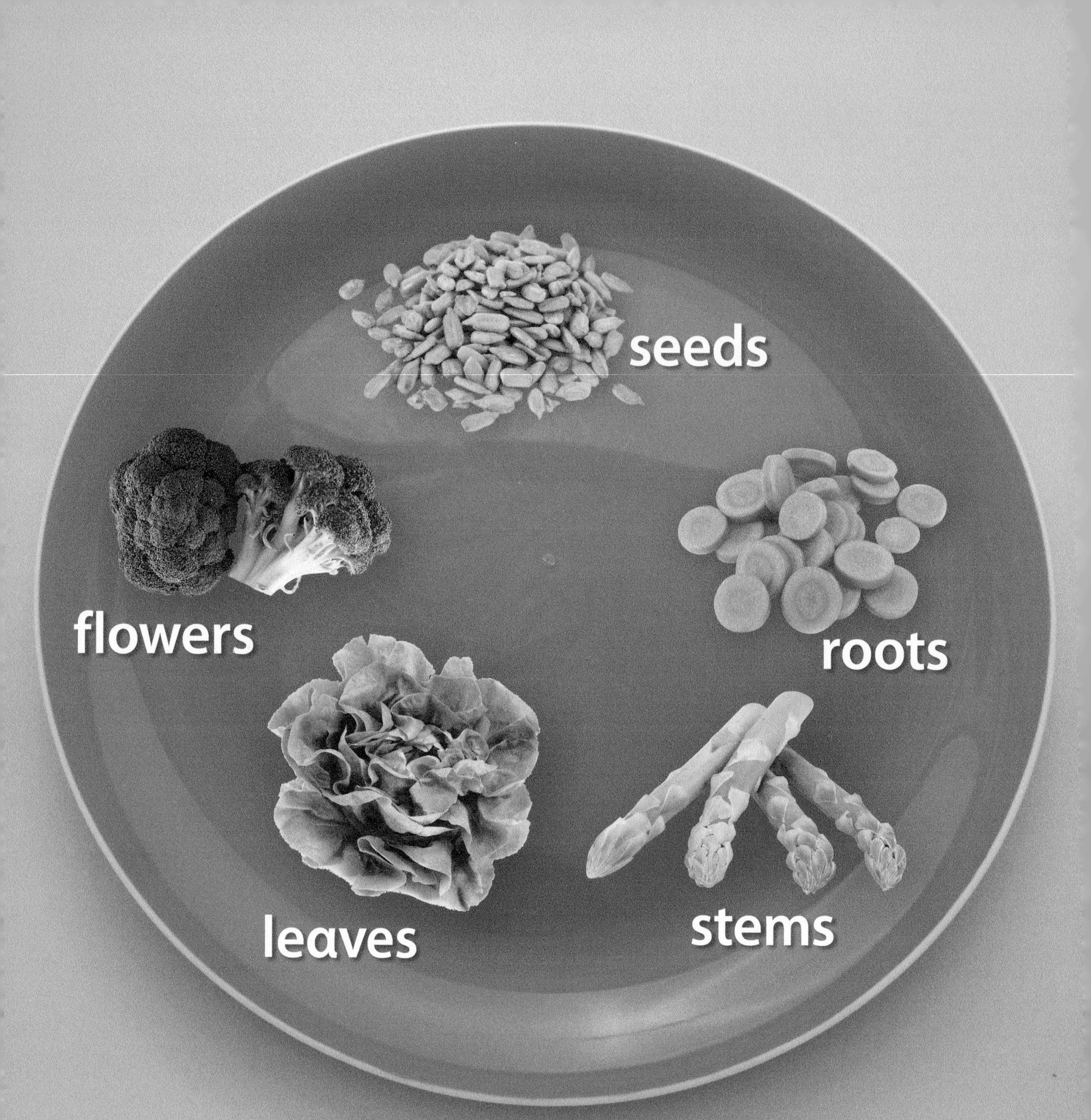
seeds
flowers
roots
leaves
stems